Efficacy of Praziquantel, the Antischistosomal Drug

Grace Naanyu Kisirkoi

Efficacy of Praziquantel, the Antischistosomal Drug

Determination of the Efficacy of Different Doses of Praziquantel (PZQ) on BALB/c mice Infected with Schistosoma mansoni

LAP LAMBERT Academic Publishing

Cover image: www.ingimage.com

Publisher:
LAP LAMBERT Academic Publishing
is a trademark of
International Book Market Service Ltd., member of OmniScriptum Publishing Group
17 Meldrum Street, Beau Bassin 71504, Mauritius

ISBN: 978-3-659-12169-2

Zugl. / Approved by: Nairobi, Jomo Kenyatta University of Agriculture and Technology, Diss., 2008

TABLE OF CONTENTS

Dedication

To my father Samson Kisirkoi, my mother Florence Kisirkoi, my sister Sarah, and my brothers Tim and Caleb

Preface

Schistosomiasis ranks second to malaria in its prevalence and virulence in tropical and subtropical countries and puts more than 600million people at risk of infection. Antischistosomal drugs include: oxamniquine, metrifonate and praziquantel (PZQ). The latter is the drug of choice for all schistosome infections. However, its standard dosage in mice (450mg/kg) does not clear the schistosomes in circulation, hence there is need to establish an effective dosage. The aim of this study was to establish an effective PZQ dosage with respect to gross pathology, histopathology and the worm counts. In this study, 24 mice (6 per group) were infected with *S. mansoni* at 4 weeks prior to treatment with different doses of PZQ (450, 600, and 900 mg/kg body weight and no drug, (control group). The following assays were carried out: gross pathology of the liver, perfusion for worm recovery and histopathology. Total worm counts and differential sex counts of worms from each infected mouse was determined manually under light microscope. The mice are not visibly affected by the different drug concentrations; also the symptoms of schistosomiasis are not apparent. However, gross pathology of the standard dosage (450mg/kg) group showed granulomas, inflamed livers and lesions. The livers of 5 of the mice treated with 900mg/kg had a slight inflammation and they had developed lesions but one had a normal liver. Three of the mice had granulomas while the rest had none. In the control groups, all the mice had inflamed livers, granulomas and 4 had lesions. Histopathology also revealed granulomas in the 450mg/kg group as well as the 600mg/kg group. Drug concentrations had significant (P = 0.013) effect on total worm counts recovered. Contrary, the drug had the same effect on both sexes of worms exposed (P = 0.511). There was also no significant difference (P = 0.470) on total worm counts between the mice sexes. The effect of drug on worms was more significant at 900 mg/Kg (P = 0.03). This study thus established that an increase in dosage of PZQ leads to fewer worm counts and less pathological effects on the definitive hosts (most importantly-man). As per this study the dosage-900mg/kg was the most effective, and doses beyond this were not investigated. The dosage groups had 25.2%, 21.4%, and 16.7% respectively (450mg/kg, 600mg/kg, and 900mg/kg) of the total worm counts, while the control group had a hefty 36.7% of all the worms recovered. Hence, with variations of the dosage of this antischistosomal drug of choice it is possible to come up with an effective dosage that clears all the schistosomes from circulation, thus stopping the infection.

1.0 INTRODUCTION

Schistosoma mansoni is a significant parasite of humans causing schistosomiasis. The disease continues to spread regardless of sustained control measures and the availability of an effective drug, praziquantel (Petney, 2001). The number of infected people is estimated to be 250 million, while those at risk of infection are approximately 600 million (WHO, 1993). The estimates of global disease burden originally produced a low ranking for schistosomiasis and produced a figure of 93 million disability adjusted life years (DALY) (Murray *et. al*, 1996; WHO 2002a). Based on recent studies it has been estimated that there is 2-15% disability in the average person with schistosomiasis, representing a 4-30 fold increase in disability weight (King *et al*, 2003). The spread and transmission of schistosomiasis has been found to be related and associated with contact with fresh water bodies.

2.0 LITERATURE REVIEW

2.1 Species of Schistosoma

The three main species of *Schistosoma* that cause human infection are: *Schistosoma mansoni*, *Schistosoma haematobium*, and *Schistosoma japonicum*. However, two other species are capable of producing disease in humans, albeit to a much lesser extent. They are *Schistosoma mekongi* and *Schistosoma intercalatum*. On rare occasions man may also be parasitised by animal schistosomes such as *S. bovis, S, curassoni, S. margrebowiei, S. mattheei, and S. rodhiani* (Sturrock 2001). *S. mansoni, S. japonicum, S. mekongi* and *S. intercalatum* cause intestinal schistosomiasis while *S. haematobium* causes urinary schistosomiasis.

2.2 Geographical Distribution of *Schistosoma mansoni*

Schistsoma mansoni is found in 54 countries including the Arabian Peninsula, Egypt, Libya, Mauritania, Somalia, Sudan, Sub-Saharan Africa, Brazil, Surinam and Venezuela

(Michael 2005). It is geographically distributed in Tropical Africa, South and Central America, Caribbean Islands and part of South West Asia. In Kenya, *S. mansoni* infections occur in Nyanza Province and in Western Province. It occurs especially in localities found along the shores of Lake Victoria e.g. Asembo (Rarieda Division). The trematode has also been found to greatly affect people, especially, school going children with proximity to water bodies such as River Kambu, hence the infection is rampant in other such areas in the country including Mwea, Coast, and Msambweni areas, Machakos, Kitui and Taita Taveta (WHO 2002a; WHO 2002b). A study in Kisumu town showed very high levels of *S. mansoni* infection among car washers working along the shores of Lake Victoria. These results suggested that rural communities along the lake might have high prevalences as well. Hence schistosomiasis has a tendency of having a focal geographical distribution in that the infection is especially rampant in isolated foci where the environment favours the intermediate host and subsequently, the definitive host (a vertebrate) is present and in contact with water bodies infested by the intermediate host (Zaris, 2007).

2.3 Hosts of *S. mansoni*

S. mansoni has a heteroxenous life cycle maintained in an intermediate host (snail) and a definitive host (a vertebrate e.g. man, mouse, baboon). The intermediate host is a fresh water snail of the genus *Biomphalaria*. Different *Biomphalaria* species transmit schistosomiasis in different localities. In Africa the most important snail host is *Biomphalaria pfeifferi* group. It is a very efficient intermediate host found in streams, seepages and a variety of man-made water bodies e.g. dams, water channels and swimming pools. Normally it occupies the central portion of Africa, with the northern limits in Ethiopia, Lake Chad and Senegal (Williams and Hunter, 1998).

The other snail groups found in Africa include:

- *Chaonomphala* group- Occurring mainly in lakes. It is the primary vector found in Lake Victoria.

- *Alexandrina* group- Occurring sporadically in North, East and Southern Africa.

- *Sudanica* group- Occurring in East and West Africa (Cook 1996).

The main snail host in South and Central America is *Biomphalaria glabrata.*

The host in which a parasite attains its sexual maturity is dubbed "definitive host" and the host in which asexual reproduction of a parasite takes place is dubbed "intermediate host".

Vertebrates would generally pass for definitive hosts of *S. mansoni*. However, man is the most important definitive host although other definitive hosts may include rodents, baboons and vervets. (Cook, 1996).

3

2.4 Life Cycle of *S. mansoni*

Man is infected during the course of activities that involve water contact with the skin. The penetrating cercaria migrates through intact human skin to dermal veins. During penetration, cercaria shed the tail and the outer glycocalyx and develops a double lipid bilayer tegument highly resistant to immune responses (shortly post-infection events). It is now a "schistosomula" (Behrman, 2005). Transformation into schistosomula (by successful penetrants) occurs within the skin. Schistosomula enter directly into the vascular system or indirectly into the lymphatics (Jordan *et al.*, 1993).

A schistosomula exits from the skin a few days later into the circulation beginning at the dermal lymphatics and venules.

Schistosomiasis

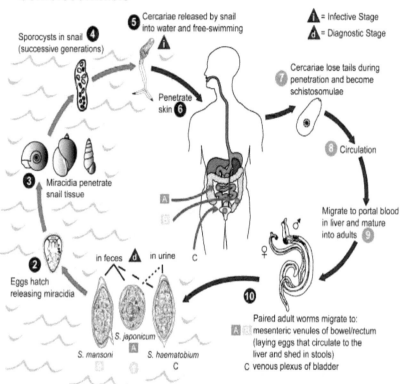

Life cycle of *Schistosoma mansoni*,

Source: Center for Disease Control

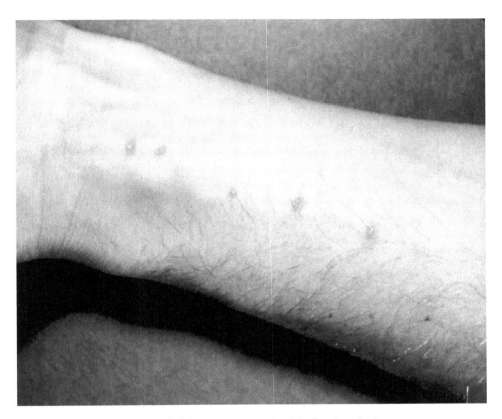

Illustration Showing Entry Point of *Schistosoma mansoni* and the Resultant Lesions,

Source: Center for Disease control

Five to seven days after penetration schistosomes migrate into the lungs then to the left side of the heart to the hepatic portal circulation (which occurs more than 15 days post-penetration). The worms start feeding on red blood cells in the liver and develop to maturity and migrate up the mesenteric veins to the gut wall. The ventral surface of the male forms a groove called the "gynaecophoric canal" in which the adult female worm resides (Zaris, 2007). This pair of worms moves against blood flow to the mesenteric circulation where the mature female deposits spined eggs on the endothelial lining of the venous capillary walls starting 4-5 weeks post-skin penetration. A proportion of the approximately 300 eggs excreted by the mature female worm per day are primarily excreted in faeces, having been deposited into the lumen of the intestines and mixing with the food bolus as it descends down the alimentary canal into the rectum. In the event that the eggs (once excreted alongside faeces) come into contact with fresh water, a larva known as a miracidium hatches. For the life cycle to continue, the miracidium must enter the appropriate hosts, which for *S. mansoni* are snails of the genus *Biomphalaria* and specifically *B. pfeifferi* in Eastern Africa. There are two generations of asexual multiplication in the snail. The miracidium transorms into a mother (primary) sporocyst containing germ balls that develop into daughter (secondary) sporocysts within a period of 1-2 weeks. When released secondary sporocysts migrate to the nutrient rich environment of digestive glands. A second phase of multiplication follows when germ balls in each secondary sporocyst mature forming cercaria, which emerge out of the snail host under, appropriate circadian and daily photophase cycles (Jourdane *et al.*, 1997). Up to 3000 cercaria per snail per day may be released and they are able to penetrate the skin of the final (definitive) host.

2.5 Disease Patterns of *S. mansoni*:

Many individuals experience no signs of the infection probably until 4-6 weeks post infection where one gets a general ill feeling. However, others especially foreigners get symptoms on the first day.

A. Within 12 hours of infection an individual may complain of a tingling sensation or a light rash, commonly called "swimmer's itch" due to irritation of the point of entrance. The rash that may develop could mimic scabies and other types of rashes. This lasts for 1-2 days. Two to ten weeks later other symptoms may occur including: fever, aching, cough diarrhoea, or gland enlargement. These symptoms can also be related to avian schistosomiasis, which does not cause any further symptoms in man (Picquet *et al.*, 1998).

B. Katayama fever (acute schistosomiasis) is another primary condition that may also develop from infection with these worms and it can be very difficult to recognize. It occurs about 3-8 weeks after the symptom free period. It appears as allergic manifestations characterized by: i) Fever, chills, cough, sweating, headache, lethargy.

ii) Eruption of pale, temporary bumps associated with severe itching (urticaria) rash, muscle and abdominal pain, patches of pneumonia, bronchospasm. iii) Splenomegally (spleen enlargement /inflammation; Guirguis, 2003).

C. Chronic schistosomiasis occurs as a result of most patients being either asymptomatic or mildly symptomatic, hence they do not usually seek medical attention and the infection therefore progresses to its chronic phase (Palaniandy, 2005). Chronic schistosomiasis is further subdivided to include: intestinal schistosomiasis and hepatosplenic schistosomiais. Chronic schistosomiasis is characterized by portal hypertension with esophageal varices and haemorrhoids causing haemorrhagic manifestations, pulmonary cor-pulmonale, immune-complex glomerulonephropathy, and transverse myelitis due to involvement of the spinal cord.

A heavy worm burden is upon a small proportion of the endemic population, thus leading to clinical complications (Palaniandy, 2005).

D.   In intestinal schistosomiasis eggs become lodged in the intestinal wall and cause an immune system reaction called a granulomatous reaction. This immune response can lead to obstruction of the colon and blood loss. The distal recto sigmoid colon is most commonly involved in intestinal schistosomiasis. Eggs are released from ova by mechanical pressure and cytolytic enzymes into gut lumen to be excreted in stool. The infected individual may have what appears to be a potbelly. If ova are retained in the intestinal tissue, eggs are disintegrated with fibrosis at the site. The mucosa of the part involved shows granulation tissue with ova, causing polyploidy hyperplasia of mucosa, which often bleeds and ulcerates. Ulcers with dense fibrosis cause narrowing of the gut lumen (Andrade and Freitas, 1991).

E.   In hepatosplenic schistosomiasis, eggs become lodged in the liver leading to high blood pressure through the liver, enlarged spleen, the build-up of fluid in the abdomen and potentially life threatening dilations or swollen areas in the oesophagus or gastro intestinal tract that can tear and bleed profusely [oesophagal varices; Andrade and Freitas, 1991]. Ova are deposited in minute vessels of the sub-mucosa. If the ova cannot be excreted through the intestine, it may be carried through portal vein to the liver producing portal tract fibrosis. Hepatosplenic schistosomiasis is characterized by reticuloendithelial cell hyperplasia and granulomas in the liver and spleen. Ova are lodged in the small hepatic portal radicles and contain a living embryo that can survive for a period of 2-3 weeks. It secretes soluble antigens and induces a granulomatous reaction. The intra hepatic portal radical is totally replaced by a granuloma that occludes the lumen. Acute endophlebitis of the intra hepatic radicles may be noted. This could finally lead to intra hepatic vascular block. The inflammations as well as destruction of the coats of blood vessels lead to thrombosis of the blood vessels. The thrombi become recanalized and the newly obtained blood vessels communicate through the walls of the vein with adjacent walls outside forming telangiectasis in the portal area. As a result of prominent vascular and fibrotic changes, the portal area is moderately broadened. The liver is grossly enlarged and slightly nodular. On cross section, portal fibrosis stands out termed as "pipe stream fibrosis". Intra hepatic portal fibrosis causes portal hypertension. Oesophageal varices complicate the picture of portal hypertension. Rapture of varices usually results in the death of patients with hepatosplenic involvement. The spleen is seen to be enlarged with fibrosis and contains gamma – gandy bodies (Andrade and Freitas, 1991; Daniel *et al.*, 2002).

For sufficient number of ova to reach the lungs, portal hypertension with collateral venous channels must be present to allow direct passage of eggs to the right side of the heart and from there to the pulmonary arterial tree. Thus cardiovascular bilharziasis almost always occurs in patients with hepatosplenic schistosomiasis. There are granulomas around ova, which may extend into the lumen of the pulmonary vessels. The inflammation process may destroy the wall of the artery. Newly formed arteries in both intra-arterial or Para- arterial granulomas form anastomes with pulmonary veins. Pulmonary obstructive arteriolitis causes pulmonary hypertension with right-sided heart failure (cor-pulmonale). Majority also have histopathologic findings consistent with a diffuse membrano-proliferative gromerulonephritis also called immune complex glomerulonephropathy of the kidney. Ectopic lesions may be observed as solitary granulomas around ova. They may involve any organ or tissue without any clinical effect. Solitary lesions are composed of numerous eggs, pseudotubercules, granulation tissue and varying amounts of fibrous tissue. Central nervous system lesions are also experienced in certain rarer cases. The effect is usually on the brain causing symptoms of cerebral irritation and spinal cord causing transverse myelitis (Palaniandy, 2005).

2.6 Drugs Used Against Schistosomiasis

Praziquantel (PZQ) is a heterocyclic pyrazino-isoquinololine, which could be used, in a single dose and it is the drug of choice in schistosomiasis treatment (Fenwick 2003). It has antischistosomal activity against all five species of the parasite capable of producing disease in man *(Schistosoma mansoni, S. haematobium, S. japonicum, S. mekongi, and S. intercalatum).* Principally praziquantel targets adult schistosomes but it is active against schistosomula in the first two days after penetration into the skin. After this time and up to about 4 weeks later the parasites lose their susceptibility and can continue development progressing to mature egg producing adult worms (Gerald and Larry, 2000). While praziquantel is safe and highly effective in curing an infected patient it does not prevent reinfection by cercaria and is therefore not an optimum treatment for people living in endemic areas. There is ongoing extensive research aimed at developing a vaccine that will prevent the parasite from completing its life cycle in man. Repeated treatment with praziquantel increases the relative effectiveness of the drug compared with a single dose and it is confirmed that immature worms are capable of surviving exposure to praziquantel hence a combined treatment with the anti-malarial drug artemether which helps to clear these immature schistosomes is recommended (Picquet *et al.*, 1998; Gryseels *et al.*, 2001). Praziquantel is administered per body weight (e.g. in man, 30-60mg per kilogram body weight). It is rapidly absorbed from the intestines and quickly metabolized. The parasites are subjected to tegumental damage; they are contracted and shifted to the liver and destroyed (Andrews, 1995).

Antimonials/Antimony compounds were the first drugs used against schistosomiasis. They were however abandoned due to the severe side effects pegged to their administration. In low doses this toxic metalloid bonds to sulphur atoms in enzymes used by the parasite and kills it without substantial harm to the host. This treatment is considered outdated and overshadowed.

Lucanthone hydrochloride was used between 1948 and the mid-sixties but was abandoned as well due to frequent side effects. These included nausea, vomiting, and occasionally diarrhoea.

Niridazole has been use since 1964. It had a mutagenic effect among other side- effects. It was also proved to be carcinogenic in mice and hamsters (Legator, *et al.*, 1995).

Oxamniquine has been used in the treatment of acute, sub- acute, chronic and complicated cases of *S. mansoni* infections with uniform good effects and side effects such as headaches, dizziness, drowsiness (Petney, 2001; Singh, 1997). However praziquantel seems to have replaced it since the 1980s.

Metrifonate is ideal for mass therapy against *S. haematobium* and may be effective for prophylaxis. It is not effective against *S. mansoni*. It is low cost and high efficacy has made it an important agent in low – income countries (Wilson, 1993).

Mirazid is a new Egyptian drug under investigation for oral treatment of schistosomiasis.
Medicinal castor oil as anti penetration agent has been experimentally shown to prevent schitosomiasis.

In addition praziquantel's effectiveness is also dependent upon the vehicle used to administer the drug e.g. cremophor/ castor oil, this makes it more effective (Petney, 2001).

Initiation of antischistosomal therapy may result in coughing and wheezing accompanied by new infiltrates on chest radiograph, as well as eosinophilia due to an immunologic response from dead worms. (Wilson, 1993)

3.0 STATEMENT OF THE PROBLEM

Praziquantel (PZQ) is the most favoured drug used for treating schistosomiasis. However, it is more effective (in normal standard doses-450mg/kg in BALB/c mice) in clearance of eggs in urine (for *S. haematobium*) or stool (for *S. mansoni)* than in clearance of the worms from systematic and pulmonary circulation. The dosage used in treatment of experimentally infected mice at the Institute of Primate Research (IPR) is 450mg/kg, which also does not completely clear the worms. The dosage in mice (450mg/kg) is apparently merely sub-curative, explained by recovery of worms after treatment in mice. Mice are used in this study after which hamsters, rabbits then non-human primates can be used as a model for humans in the search for an efficient dose of praziquantel.

4.0 JUSTIFICATION

PZQ in its standard dosage of 450mg/Kg in mice as routinely used at the IPR does not absolutely annihilate the schistosomes in circulation. Hence there is need to establish a new and workable as well as effective dosage to be used in treatment of schistosomiasis. It is therefore necessary to test the efficacy of other higher doses in mice.

5.0 OBJECTIVES
5.1 General Objective

To determine the efficacy of different doses of praziquantel on BALB/c mice experimentally infected

with *S. mansoni.*

5.2 Specific Objectives

1) To determine the effect of different doses of PZQ on worm counts in BALB/c mice experimentally infected with *S. mansoni.*

2) To determine the effect of different doses of PZQ with regards to gross pathology in BALB/c mice experimentally infected with *S. mansoni.*

3) To determine the effect of different doses of PZQ on the histopathology of BALB/c mice experimentally infected with *S. mansoni.*

6.0 NULL HYPOTHESIS

Increasing the dosage of praziquantel in BALB/c mice infected with *S. mansoni* does not improve the

efficacy of the drug.

3.0 MATERIALS AND METHODS
3.1 S. *Mansoni* parasites
A strain of *S. mansoni* originally obtained from humans, and maintained in olive baboon

(*Papio anubis*) at the Institute of Primate Research (IPR) was used for all the work for hatching

miracidia to infect mice.

3.2 Hosts

3.2. 1 Definitive host
BALB/c mice are permissive hosts of *S. mansoni*. They were bred at the Animal Resources Department

at the IPR. They were fed on nutrient pellets (Laboratory chow, Unga feeds ® Co.) and supplemented

with carrots and kale leaves. Water was supplied ad libitum (orally).

3.2.2 Intermediate host
Biomphalaria pfeifferi snails are the intermediate host of *S.mansoni*. They were collected from

Kakuyuni River in Kangundo in Machakos District. The snails were scooped out of water using a scoop

with a long wooden handle.

3.2.2.1 Maintenance of Snails in the Malacology Laboratory
The snails collected from Kangundo were carried in plastic containers lined with damp cotton wool and

transported to the snail laboratory at the IPR where they were screened for schistosomes under strong

light (100 watts) for three hours for five consecutive weeks. Those that were negative were housed in the

temperature controlled (25-27^{0}C) snail room. Plastic tanks were washed thoroughly with tap water. Sand

and gravel sterilized by heating at 150^{0}C for 12 hours was cooled and layered in plastic tanks. Plastic

tanks were ¾ filled with tap water (chlorine free water from the IPR well). The screened snails were transferred into the tanks for maintenance. The water was changed twice a week. Lettuce dried in an oven at 40^0C for about half an hour was added to feed the snails as described by Yole, *et al.*, (1996).

3.2.2.2 Harvest of *S. mansoni* Eggs from Baboon's Faeces and Hatching of Miracidia

A faecal sample from infected baboons was collected, eggs harvested and miracidia hatched. The faecal sample was placed in a plastic beaker and one litre of IPR well water added to completely cover the sample. The mixture was stirred with a wooden spatula to obtain a thin suspension of saline and faecal material. The sample was sieved through two sieves (mesh size 600μm and 250μm) into a collecting tray and the filtrate transferred into urine jars. The urine jars were filled with saline and left 30 minutes in the dark. The supernatant was poured off without disturbing the sediments. The urine jars were filled again with IPR well water and the above procedure was repeated three times until the supernatant was clear. The sediment was transferred into glass petri dishes. Fresh water was added gently to cover the sediment. The petri dishes were placed under a 100 watts lamp for 30 minutes, as described by the method of Yole *et al.*, (1996). Emergence of miracidia was determined using the dissecting microscope with 10-40x magnification.

3.2.2.3 Snails Infection

Snails were infected using infecting wells. A Pasteur pipette with a rubber bulb (teat) was used to pick 3-6 miracidia from the glass petri dish (in procedure 2.4 above) under the dissecting microscope (10-40x magnification). The miracidia were dispensed into each well in a 24 well culture plate. Snails were transferred individually into these wells and the plates covered to prevent the snails from crawling out. The set up was left for 30 minutes to allow penetration after which snails were transferred into newly prepared aquarium tanks. The patent period for *S. mansoni* is 5 weeks. At 4 weeks post infection (p.i.),

17

the tanks were covered with dark clothes to prevent trickle shedding of cercaria. These are as described by Yole *et al.*, (1996).

3.2.2.4 Shedding of Infected Snails
Snails were artificially stimulated to release cercaria. Five weeks p.i, the snails were removed from the tanks in the dark and placed into 100 ml beakers with 20 ml of snail water. They were exposed to light (100watts bulb shielded with glass to safeguard the snails from heat) for 2-3 hours to shed the cercaria. Cercarial concentration in the suspension was estimated by counting three 50μl aliquotes under the dissecting microscope. An average of cercarial number was taken and used to calculate the concentration per ml. This was used to determine the volume of the suspension that would have 250 cercaria to infect one mouse according to the procedure described by Yole *et al.*, (1996).

3.2.2.5 Infecting Mice with the Parasite
One-milliliter syringe was used to anaesthetize mice intraperitoneally using the following procedure {described by Yole *et.al.*, (1996)}. Mice were anaesthetized with ketamine/xylazine mixture (20:1, made by adding 0.5 ml of xylazine to 10ml of ketamine). Anaesthesia was delivered as 0.02 ml per 30g mouse body weight. Once unconscious the mice were shaven in the stomach and arranged in a wooden infecting rack. Cotton wool was dipped in water and used to clean and wet the shaven area to allow easy penetration of the cercaria. One-centimeter diameter metal rings were placed on the shaven area. One ml micropipette was used to dispense the appropriate volume of the crecarial suspension containing 250 cercaria (as described in 7.6 above) into the metal ring. Cercaria were given 30 minutes to penetrate.

3.3 Administering Varied Doses of PZQ

The standard dosage for treatment of mice infected with *S. mansoni* at IPR is 450mg/kg.

1) Four hundred and fifty milligrams (mg) of the drug was given per 1kg (1000g). How many mg of the drug should be given for 30g?

30 x 450/1000 = 13.5mg (per mouse) suspended in 50μl of water.

There were 6 mice per group but the suspension was prepared to suffice 10mice such that it takes care of spillages and any such losses hence:

If one mouse was to get 13.5mg of the drug, 10 mice would get – 13.5 x 10 = 135mg suspended in 50μl x10 = 500 μl.

However the sectioned PZQ tablet was 600mg and each of the 4 sections are equal i. e. 150mg. For ease of dividing the tablet 150 mg was used (rather than 135mg).

If 135mg dissolved in 500μl of water 500mg would dissolve in – 150mg x 500/135mg = 555.5μl of distilled water.

2)   For 600mg/kg:

One mouse weighed 30g (0.03kg), hence the drug given per mouse – 0.03kg x 600/1kg = 18mg per mouse, 180mg for 10mice. 180mg of the drug would dissolve in 500μl of water to make a suspension equivalent to the dose of 600mg/ Kg.

However 225 mg was used due to the ease of sectioning the drug i. e. 1/4+1/8 of the tablet = 75mg +150mg.

If 180mg dissolved in 500μl of water, 225mg would dissolve in = 225mg x 500μl/180mg = 625 μl of distilled water.

3) For 900mg/kg, one mouse gets – 0.03kg x 900mg/Kg = 27mg, for 10 mice = 27 x10 =270mg would

19

dissolve in 500μl of water. However 300mg was sectioned from the tablet with more ease.

If 270mg dissolved in 500μl of distilled water 300mg would dissolve in –

300mg x 500μl/270mg = 555. 5μl.

The mice infected with *S. mansoni* were divided into 3 groups: 900, 600 and 450mg/kg groups. Each of the six mice in the respective groups received two doses (two days apart) of praziquantel/Biltricide® (Bayer, Germany) per 1kg body weight. The PZQ tablets were crashed in the specified quantities using a roller under a folded paper. The powder was then suspended in the specified volume in single distilled water. Using a sterile 200μl micropipette, the PZQ was slowly dispensed into the mouse's mouth using a micropipette with a yellow tip (200μl tip). There was an untreated control.

3.4 Perfusion

Mice were perfused at week 6 post infection to recover the adult worms. Each mouse was anaesthetized with ketamine/xylazine mixture (as described in 7.7 above), and the abdominal and thoracic cavities were opened up. The hepatic portal vein was incised.

Perfusion needle containing perfusion fluid (p.f. – 0.85% sodium chloride and 1.5% sodium nitrate) was inserted in the left ventricle of the heart. Perfusion was carried out until the liver and mesenteries become clear. The perfusate was collected in a large glass petri dish and transferred into urine jars for worms to sediment.

Livers were removed and stored in 100% buffered formalin for histopathological processing.

3.5 Adult Worm Recovery

Worms were recovered using the following method {of Yole *et al.*, (1996)}. The perfusate containing the recovered worms in the urine jars was topped up with phosphate buffered saline (PBS). Once the worms sedimented, the supernatant was sucked out. The settling procedure was repeated 3 times to clear the supernatant. The worms were placed in a Petri dish containing PBS and then counted using 10-40x-eye piece of a dissecting microscope. The means and standard errors of adult worms recovered from each group were calculated.

3.6 Pathological Examination

3.6.1 Gross pathology

The livers were examined for colour, size, adhesions, and presence of granulomas. Granulomas appear as numerous raised pinhead sized white foci distributed over the surface of the liver lobes.

3.6.2 Histopathology

The fixed livers were removed from 10% formalin and transferred into wooden boards in a hood. Sharp forceps were used to cut small pieces. One piece from each sample was transferred into tissue cassette and immersed into 80%, 95% and 100% ethanol respectively to achieve optimum dehydration. They were cleared in toluene, infiltrated in hot paraffin and embedded in tissue embedding paraffin wax (Sherwood Medical Company U.S.A.) on embedding blocks. The tissues were sectioned serially at 6 microns using a rotary microtone (Leitz Germany). The thin tissue sections were mounted on microscope slides and stained with haematoxylin and eosin. They were observed under the light microscope and only granulomas containing an ovum at the center were enumerated and measured using calibrated ocular micrometer. Granuloma size was measured based on the vertical and horizontal diameters. The average of vertical and horizontal diameter was taken to be the granuloma diameter.

4.0 RESULTS

After completion of the study, the results were compiled and they contained the worm counts after perfusion, gross pathology and histopathology results.

The **table 1** contains the results of worm counts done manually under a light microscope after perfusion of the mice, which had been infected with *Schistosoma mansoni* and treated with PZQ. The control group to which no drug was administered had the highest number of worms recovered from its circulation. The numbers decreased steadily with increasing drug concentration. There was no significant effect of the drug concentration on the male versus the female worms and on their development ($P = 0.511$). The drug had a significant effect on the total worm counts ($P = 0.013$).

Mice Groups	Female Worms	Male Worms	Mature Worms	Stunted Worms	Total
450mg/kg	29	59	88	5	181
600mg/kg	29	54	85	6	154
900mg/kg	22	38	60	0	120
0mg/kg(Control)	45	109	154	7	264

Table 1: Worms recovered in mice groups administered with varied doses of PZQ drug.

Table 2 shows the occurrence of granulomas, adhesions, and inflammation on the mice livers following gross pathology.

Inflammations and adhesions of the mice livers were observed in all the mice groups. These are the visible signs of the disease that were otherwise not apparent before cutting the mice open. Granulomas were not observed in the 900mg/kg group, hence depicting a reduction in the visible effects of the disease with increase in dosage. The drug had no significant (P = 0.470) effect (selectively) on any of either the male or the female mice.

Dosages	Mice gender	Granuloma	Adhesions	Inflammation
450 mg/kg	Male	_	+	+
	Female	+	+	+
600 mg/kg	Male	+	+	+
	Female	+	+	+
900 mg/kg	Male	_	+	+
	Female	_	+	+
0mg/kg (Control)	Male	+	+	+
	Female	+	+	+

KEY: (-)   Negative;
        (+)   Positive

Table 2: Occurance of granulomas adhesions, and inflammation on the mice livers following gross pathology: Occurance of granulomas, adhesions, and inflammation on the mice livers following gross pathology

The worms recovered from the experimentally infected mice with respect to the drug concentration used to treat them are shown in **figure 1**.

The group to which no treatment (control) was given had the highest mean number of worms recovered. The group to which the highest drug concentration was administered had the least mean number of worm counts; hence worm counts are decreasing with increase in drug concentration. There was a significant effect of the drug concentration (900mg/kg) on worm counts ($P<0.05$ i.e. $P=0.03$) this implies that the drug praziquantel at a concentration of 900mg/kg significantly destroyed worms in blood circulation of the mouse model in experimental use. The rest of the concentrations had insignificant effect ($P>0.05$). Hence, with increase in drug concentration there is a marked and heightening effect on the schistosome worms in blood circulation.

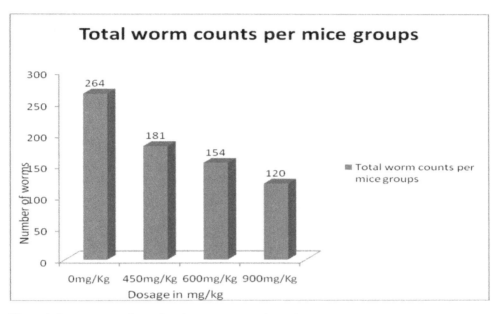

Figure 1: Drug concentration and total worms recovered per mice group

Histopathology results

After histopathological processing of the mice livers, they were stained using Hematoxylin and Eosin so as to observe the liver cells and their pathology.

Plate 1 shows the *Schistosoma mansoni* egg marked using a black line across its width. It appears centrally located on the plate surrounded by other cells of the immune system.

Plate 2 illustrates a contrast between the normal liver cells (hepatocytes) on the left side of the plate and the immune cells on the right side of the plate. This occurs in the liver tissues where the egg laid by adult schistosome worms become lodged and eventually calcify. This foreign egg is detected by the host's immune system, which chemotactically migrate to the foreign antigen to counter the invasion.

Plate 3 shows how the immune effector cells surround the *Schistosoma mansoni* egg forming up a granuloma. The longest distance marked by a line on the plate depicts the length of the granuloma whereas the marked shortest distance across depicts its width. These two distances are averaged to give the granuloma's size. The mean size of the granulomas was 24.7μm.

Histopathology Results

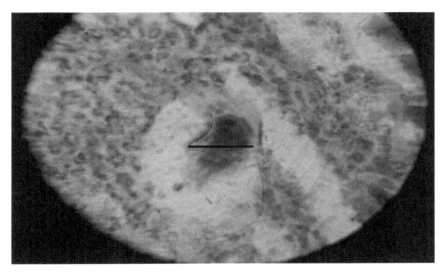

Plate 1: The marked *Schistosoma mansoni* egg

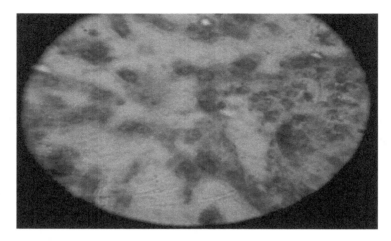

Plate 2: Contrast between normal hepatocytes (left side) and immune cells making up the granuloma (right side)

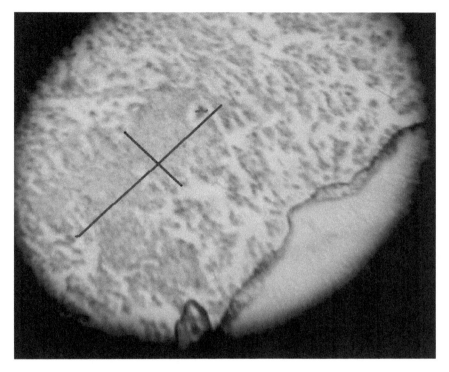

Figure 2:
Plate 4: Granuloma measurements showing length (longest distance along) and width (shortest distance across)

5.0 DISCUSSION

Higher doses (900mg/kg and 600mg/kg) of praziquantel as was used in this study were more effective than the standard dose used routinely on the mouse model. This study has proved the necessity of altering the drug regimens in current use as these haemoparasites cause a lasting endemicity in susceptible populations with repeated recurrence of sporadic cases vindicated by worms recovered from the mice's circulation despite drug administration, (**table 1** and **figure 1**).

Eggs are also seen to be embolized in hepatic venules with formation of granulomas. Granulomas are composed of aggregation of mononuclear cells, neutrophils, lymphocytes, plasma cells and fibroblasts. Since their sizes were being measured as the granulomas were being observed under the eye piece stage meter slide, there was need to convert these calibrations to those of the stage micrometer's.

Hence:

Stage micrometer slide-[100 subdivisions x 0.01= 1mm]

25 small divisions of the eye piece stage meter slide = 20 small divisions of the stage micrometer.

Thus:

1 small division of the eye piece stage meter slide = 0.8 small divisions of the stage micrometer slide i.e. 0.008mm. This was factored in all the calculations of the granuloma sizes.

These granulomas were rampant in the control group, and their mean size was 24.7µm. This was so because the control group was infected but was not treated with any dose of drug (untreated positive control).

Praziquantel (2-cyclohexylcarbonyl-1,2,3,6,7,11b-hexahydro-4H-pyrazino[2,1-a]isoquinolin-4-one) was synthesized in the 1970s (Seubert *et al.*, 2002). While undergoing initial veterinary screening it showed efficacy against cestodes (Groll, 1984; Thomas and Gönnert, 1977). With regard to schistosomes, efficacy was first demonstrated against *S. mansoni* in different host animals (Gönnert and Andrews, 1977 Pellegrino *et al.*, 1977). These findings were confirmed for all other schistosome parasites pathogenic to humans (Andrews, 1981; James *et al.*, 1977). The stage-specific susceptibility of a Puerto Rican strain of *S. mansoni* harbored in mice treated with three oral doses of praziquantel on alternate days has been studied. Although only few replicates were used, these studies revealed that the invasive stages (cercariae and very young schistosomula) and the mature worms were affected more than the liver stages of the parasites. These findings were confirmed for *S. japonicum* (Xiao *et al.*, 1987).

6.0 CONCLUSION

Alterations of the drug dose administered against BALB/c mice experimentally infected with *Schistosoma mansoni* has direct effect as depicted by histopathology and gross pathology. Worm counts recovered after the treatment steadily declined with increase in the dose of the drug. With increase in the dose of PZQ, there is a steady rise in efficacy of the drug with respect to reduction of worm burdens in circulation, reduction in granulomas as observed regarding the disease pathology.

7.0 RECOMMENDATIONS

✓ This study can be further progressed by trying other dose regimens using combinations of those dosages used in this study e.g. 900mg/kg followed by smaller doses e.g. 450mg/kg in the mouse model.

✓ Varied replications of treatments may be tried e.g. administering three dose replications with diminishing doses instead of the same quantity.

✓ The study may be further carried out on the models that are closer to man e.g. non-human primates, to ensure clear elucidations of the outcome before clinical trials are done.

✓ Another wider approach may be taken, including a study to establish whether these new doses are safe for use and at what maximum level (toxicity) e.g. testing for drug potential of teratogenicity, carcinogenicity, mutagenicity or effect of sterility.

Acknowledgements

I heartily acknowledge my supervisor Prof. Dorcas Yole for her diligent guidance and support at every stage of the project, for her motivating words of wisdom, always upholding integrity with deep sincerity and propelling me to higher heights than I knew I was capable of. This is perfectly complemented by Dr. Helen Kutima, whose persistent concern in my work brought out intrinsic details that required special attention. The Institute of Primate Research (IPR) and its esteemed personnel were a memorable stepping stone, warmly offering its premises and resources to make my project a success, Collins Kisara was of exceptional help during my toiling moments. My father Samson Kisirkoi, My mother Florence Kisirkoi, My sister Sarah Naserian, My brothers Timothy Lemayian and Caleb Lekishon, and my niece Joy Enkirotet; this work would never have been what it is without your love and support. I thank my friend Winifred Selle for her critical eyes that were ever willing to proofread my write-up.

Above all I thank God for He is the one who gave me all what it took to actualize this work.

Thank you heartily,

Grace.

REFERENCES

Andrade, Z. A. and Freitas, L.A.R. (1991), Hyperplasia of ElasticTissue in Hepatic   Schistosomal Fibrosis *Memorias do Oswaldo Cruz* 86, 447-56.

Andrews, P. (1995) Praziquantel: Mechanisms of Antischistosomal Activity. *Pharmacological therapy*29: 129-156.

Andrews, P. 1981. A summary of the efficacy of praziquantel against schistosomes in animal experiments and notes on its mode of action. Arzneimittelforschung **31:**538-541.

Behrman, A. M. (2005), Schitosomiasis Medicine Specialties. Emergency Medicine. Infectious Diseases.

Chitsulo, L., Engels, D., Montresor, A. and Savioli, L. (2000).The Global Status of Schistosomiasis and its Control. *Acta Tropical* No. 77:41-51.

Cook G. (1996) Manson's Tropical Disease International Student Edition W. B. Saunders Bath Press Great Britain.

Daniel, G. C, Margaret, T. W, George, L. F, Jr. and Secor, W. E. (2002) Idiotypes expressed early in experimental Schitosoma mansoni. Infection predict clinical outcomes of chronic Disease. *The journal of experimental medicine,* Vol 195 No. 9 1223-1228.

De Vlas, S. J., Gryseels, B., Van Oortmarssen, G. J., Polderman, A. M. and Habbema, J. D. (1992). A model for variations in single and repeated egg counts in *Schistosoma mansoni* infections. *Parasitology* No. 104: 452-460.

Fenwick, A., Savioli, L., Engels, D., Bergquist, N. R. and Todd, M. H. (2003).Drugs for the control of parasitic diseases: Current Status and Development in Schistosomiasis. *Trends in parasitology* No. 19: 509-515.

Gerald, D. S. and Larry S.R. (2000), *Foundation of Parasitology*; Larry S. R. and John

    J. Jr 6th Edition Mc Graw Hill.

Gönnert, R., and P. Andrews. 1977. Praziquantel, a new broad-spectrum antischistosomal agent. Z.

    Parasitenkd. **52:**129-150.

Groll, E. 1984. Praziquantel. Adv. Pharmacol. Chemother. **20:**219-238.

Gryseels B. (2001) Are Poor Responses to Praziquantel for the Treatment of *Schistosoma mansoni* Due

    to Resistance? An Overview of Infection. *American Journal*

    of Tropical Medicine and Hygiene No. 92: 90-93.

James, C., G. Webbe, and G. S. Nelson. 1977. The susceptibility to praziquantel of *Schistosoma*

    haematobium in the baboon (*Papio anubis*) and of *S. japonicum* in the vervet monkey

    (*Cercopithecus aethiops*). Z. Parasitenkd. **52:**179-194.

Jordan, P.,Webbe, G. and Sturrock, R. F. (eds) (1993). *Human schistosomiasis, CAB international.*

Jourdane, J., and Theron, A. (1997).Larval Development: Eggs to Cercariae. In the Biology of

    Schistosomes: From Genes to Llatrines. (ed. Rollinson, D. and Simpson, A. J.),. academic press,

    London. pp 83-114.

King C.H., Dickman, K. and Tisch,D.J., (2003) ,Regauging the Cost of Schistosomiasis: a Meta-

    analysis in Endemic Populations. *American Journal of Tropical Medicine and Hygiene* No. 69:513.

Legator , M. S., Connor, T. H. and Stoeckel, M. (1995). Detection of Mutagenic Activity of

    Metronidazole and Niridazole in Body Fluids of Humans and Mice Science 188,1118.

        Michael, D. N. (July 25, 2005) Schitosomiasis eMedicine Specialties Pediatrics

Parasitology

Palaniandy, K. (2005) Schistosomiasis eMedicine Specialities. Medicine, Obstaetrics, Gynaecology, Psychiatry, and Surgery. *Infectious Diseases*.

Murray, C. J. L. and Lopez, A. D. (1996). *The Global Burden of Disease*. Cambridge, M.A.: Harvard University Press.

Pellegrino, J., F. F. Lima-Costa, M. A. Carlos, and R. T. Mello. 1977. Experimental chemotherapy of schistosomiasis mansoni. XIII. Activity of praziquantel, an isoquinoline-pyrazino derivative, on mice, hamsters and Cebus monkeys. Z. Parasitenkd. **52:**151-168.

Petney T. N. (2001) Environmental, Cultural and social changes and their influence on parasite infections. *International Journal of Parasitology* No. 31: 919-932.

Picquet M., et al. (1998). Efficacy of praziquantel against *S. mansoni* in Nothern Senegal. Trans royal society Journal of tropical medicine and hygiene 92: 90-93.

Sabah, A. A., C. Fletcher, G. Webbe, and M. J. Doenhoff. 1986. *Schistosoma mansoni*: chemotherapy of infections of different ages. Exp. Parasitol. **61:**294-303.

Savioli, L. (2002), Schistosomiasis and snail transmitted helminth infections; forging control efforts.*Trans Royal Society of Tropical Medicine and Hygiene*96.

Seubert, J., R. Pohlke, and F. Loebich. 1977. Synthesis and properties of praziquantel, a novel broad spectrum anthelmintic with excellent activity against schistosomes and cestodes. Experientia **33:**1036-1037.

Singh B. (1997) Molecular Methods for Diagnosis and Epidemiologic Studies of Parasitic Infections, *International Journal of Parasitology* No. 27: 1135-1145.

Sturrock, R. F. (2001). The Schistosomes and their Intermediate Hosts. *In Schistosomiasis* (Mahmoud A. A. F. Ed.), Imperial College Press, 2001 p7

28. **Thomas, H., and R. Gönnert.** 1977. The efficacy of praziquantel against cestodes in animals. Z. Parasitenkd. **52:**117-127.

Van Der Werf, M. J., De Vlas, S.J., Brooker, S., Looman,C.W.,Nagelkerke,N,J.,Habbemma,J.D. and Engels,D. (2003).Quantification of clinical morbidity associated with Schistosoma infection in Sub-Saharan Africa. *Acta Tropical Journal*

WHO (2002a). Prevention and control of Schistosomiasis and soil transmitted helminths. Report of a WHO expert committee.WHO.Technical Report Series 912, WHO Geneva.

WHO (2002b). Schistosomiasis: Distribution, Life Cycle, Pathology, Diagnosis and Control WHO Factsheet 117 : 1-120.

WHO (2006) Schitosomiasis and Soil Transmitted Helminths Country Profile.

Williams S. N. and Hunter, P. J. (1998). The distribution of *Bulinus* and *Biomphalaria* in Khartoum and Blue Nile Provinces, Sudan. Bulletin of WHO 39 948-954

Wilson R. A. (1993) Immunity and Immunoregulation in Helminth Infection. *Current Opinions in Immunology Journal* No. 5: 538-547.

Xiao, S. H., W. J. Yue, Y. Q. Yang, and J. Q. You. 1987. Susceptibility of *Schistosoma japonicum* to different developmental stages to praziquantel. Chin. Med. J. **100:**759-768.

Yole, D. S., Pemberton, R., Reid, G.D. and Wilson, R. A. (1996). Protective Immunity to I Induced in the Olive Baboon *Papio anubis* by the Irradiated Cercaria Vaccine. *Parasitology, 112*: 37- 46

Zaris, E. K. (2007) John Hopkins Microbiology Newsletter Vol 26, No. 10.